natural diversity of the
CAPE PENINSULA

To my wife Béatrice, for her love and support

natural diversity of the
CAPE PENINSULA

Stephan Wolfart

Struik Publishers

(a division of New Holland Publishing (South Africa) (Pty) Ltd)

80 McKenzie Street

Cape Town 8001

website: *www.struik.co.za*

First published in 2001

10 9 8 7 6 5 4 3 2 1

Publishing manager: Pippa Parker

Managing editor: Helen de Villiers

Editor: Jeanne Hromník

Designer: Janice Evans

ISBN 1 86872 591 X

Reproduction by Hirt & Carter Cape (Pty) Ltd

Printed and bound by Sing Cheong Printing Company

Limited, Hong Kong

CONTENTS

AUTHOR'S NOTE

When I arrived in Cape Town in 1993, my interest was instantly captured by the beauty of its setting. Amazingly, a 10-minute drive from the city centre took one into pristine natural fynbos in the most species-rich floral kingdom on earth. In Europe, we know proteas as expensive floral jewels in a vase, but here they were in all sizes, shapes and colours and in landscapes that took away one's breath.

Almost 300 sunny days a year lend themselves to photographing this diverse and interesting environment and the life that it both reveals and hides. A large number of wild creatures manage to coexist here with humans. Since my arrival in Cape Town, I have succeeded in seeing many of them, but I have often had to be content with footprints, leftovers of meals, quills and other signs of their presence.

Visits to wetlands at Strandfontein or Rondevlei Nature Reserve have never failed to produce something exciting throughout the year. A special thrill has been the appearance of rare birds from distant continents – an American Sheathbill (on the right), an American Purple Gallinule or a Whitetailed Tropicbird. Other creatures fairly well habituated to humans, but still unseen by most people, have made homes in and around houses on the Peninsula – to the dismay of many householders. The proximity of man and nature on the Peninsula, nevertheless, is something to be treasured.

Searching for, watching and shooting nature and wildlife (with a camera, of course!) is great fun, though much time and patience is required. An extensive network of trails and paths in this part of Africa allows access to many interesting areas with good vantage points and bird hides. My vehicle also serves me sometimes as a hide. Gardens and parks are rewarding and, quite often, I can watch Peregrine Falcons and even Black Eagles from the comfort of my balcony in central Cape Town. A Spotted Eagle Owl even uses the railing from time to time as a roost from which to hoot.

For the photographs in this book, I used only Nikon cameras and lenses ranging from 24 mm to 600 mm (including micro lenses) and an array of extra equipment. Fuji Provia and Sensia are superb for sharpness, colour rendition and contrast, and most of the images were taken on these emulsions. I try to avoid lens filters entirely and choose, instead, light conditions that can be used to advantage. I rarely use flash, except at night and, occasionally, to fill-in

shaded areas. Apart from frame cropping, I have not altered any of the images, electronically or otherwise.

Most of the photographs in this book were taken from easily accessible trails and places open to the public. However, I required special permission, on occasion, for access and for publication. In this regard, I wish to thank Howard Langley, managing director of the Cape Peninsula National Park. My thanks also to my friend Dave Hierons, formerly managing director at La Farge Quarries, Tygerberg, to Blouberg Municipality (for access to Milnerton Racecourse), to Ecosense and Kenilworth Turf Club (for access to Kenilworth Racecourse) – and to all those friendly Capetonians who allowed me to explore their lands and property.

I am grateful to Clifford Dorse for his friendship as well as his willingness to share his great expertise on environmental issues. Some very valuable advice was given to me, meanwhile, by the field rangers at Cape Point and at Rondevlei Nature Reserve, with Dalton Gibbs at the helm. I am greatly indebted to several experts for identifying animal and plant species: Marius Burger, Dr Simon van Noort, Dr Dee Snyman, Dr John Manning and Robin Jangle.

It was a pleasure to work with the staff of Struik Publishers. I am particularly grateful to Pippa Parker and Janice Evans for their enthusiasm and their determination to make this project a success.

My last and biggest thank you goes to my wife, Béatrice, who encouraged me in my efforts and assisted me wherever possible.

INTRODUCTION

The Cape Peninsula is the south-westerly tip of Africa. It lies about half way between the equator and the land mass of Antarctica, close to where the Atlantic and Indian Oceans meet. It is a unique, complex and delicate environment with a rugged topography, specialized vegetation, and a human presence that places large demands on it.

Extending for about 60 km from north to south, this narrow finger-like promontory consists mainly of a long mountainous spine, accessible to nature lovers through the many tracks and paths that criss-cross the mountain slopes. The most celebrated of its mountains is Table Mountain, the flat-topped bulk of which presides over the busy city of Cape Town and marks the Peninsula's northern end. Though the landscape becomes less rugged in the south and there are even some undulating plains towards Cape Point at the southern extremity, the Peninsula is predominantly mountainous.

Where the mainland begins in the north and east, the landscape changes suddenly and mountain heights give way to an extensive plain that stretches eastwards, appropriately called the Cape Flats. Here, a number of natural estuaries and lakes (vleis) hold their own amid extensive urban development. The coast around the Peninsula is variable, alternating between rocky shore, curving sweeps of sandy beach and precipitous cliffs.

Although the region is classified botanically according to vegetation types, we have chosen to focus, rather, on seven habitats that coexist in this small area. They lie between the two extremes of ocean shore and the afromontane forest found on the eastern and southern slopes of Table Mountain. In between are narrow belts of coastal strandveld, scattered streams, lakes and wetlands, small enclaves

View of Hout Bay and the Atlantic Ocean from Judas Peak.

Protea repens in Nursery Ravine, above Kirstenbosch.

African Spoonbill, Rondevlei.

Malachite Kingfisher, Rondevlei.

six floral kingdoms. The Cape Floral Kingdom contains more than 8 500 plant species in an area that is 0,2% or less of the northern hemisphere's Boreal Kingdom. The Cape Peninsula supports about 2 500 plant species, of which about 1 500 are to be found on Table Mountain alone. Of these, almost 150 are endemic, to be found growing in nature only in the area of the Peninsula.

Most of this amazing diversity is found in a deceptively dull looking vegetation type known as fynbos (fine-leaved bush), the dominant vegetation in the Cape Floral Kingdom and in the larger fynbos biome (a botanical concept referring to a community of plants defined in terms of climate and main vegetation type). Afromontane forest and renosterveld, which is similar in appearance to fynbos, are two distinct vegetation types within the region.

Despite the destructive systems imposed by humans, including the introduction of alien animals and plants that flourish at the expense of the indigenous species, we have a unique and surprising situation at the Cape Peninsula. The recently proclaimed Cape Peninsula National Park (encompassing the mountains, including Table Mountain, and natural areas of the whole Peninsula from Signal Hill to Cape Point) is one of the most important conservation areas in South Africa. Its status as part of a proposed UNESCO World Heritage site is pending.

Ironically, the existence of such a conserved area here is ultimately due to the poorness of its soils, in which fynbos thrives. Generally speaking, the Peninsula does not support agriculture or livestock and, thus, has limited appeal for farmers. This has been its salvation. However, in places like the

of lowland fynbos, patches of renosterveld on the nutrient-rich clay soils on the lower mountain slopes and the predominant habitat of the region: astonishingly diverse mountain fynbos.

Known botanically as one of the earth's richest places (it has the highest concentration of plant species in the world), the Cape Peninsula falls within the Cape Floral Kingdom, the smallest of the world's

Vineyards fronting fields in Tygerberg, once the domain of renosterveld and home of large mammals like this red hartebeest.

Rugged mountain slopes above Chapman's Peak, descending into a wooded kloof.

The pushing upwards and folding of these massive layers of sedimented sand, silt and clay into almost vertical folds is dated to about 250 million years ago. What we know today as the Cape Fold Mountains, including Table Mountain, are the weathered and worn remains of this giant mountain range. Erosion over millions of years of the nutrient-poor quartzite of which they are mostly composed (overlying the earlier layers of dark-grey Malmesbury shales and light-grey granite) has produced much of the nutrient-poor sandy soil characteristic of the Peninsula and its surroundings. On the top of Table Mountain, today, ongoing weathering is again bringing to light the small vein-quartz pebbles that were deposited millions of years ago.

Geology, topography and climate have combined in the Cape Peninsula to produce a striking range of habitats and exceptional botanical diversity. The gradual, undisturbed evolution of plants and animals has been fostered by the absence of major

Constantia Valley, nutrient-rich granite and clay-based soils support prime vineyards, and little remains of the indigenous vegetation.

The Peninsula's environment, its soils included, are the end result of the huge geologic movements and the sculpting forces of weathering and erosion that have shaped this little piece of land – a grand equation of cause and effect. At a time (about 500 million years ago) when the present continents had not yet formed and Africa was part of the megacontinent named Gondwanaland, the area now called the Cape was part of a large flat plain associated with the inland Gondwana Sea. To the layers of granite and ancient Malmesbury shale that composed this plain, inland rivers and wind erosion added deposits of sand and very hard vein-quartz pebbles. Over millions of years, these deposits turned into the sturdy rock known today as Table Mountain sandstone.

Another tablecloth above Cape Town – on Lion's Head.

geological disturbance over a very long period of time. It is evident that plants have evolved in a variety of ways to cope with marked seasonal variations, broken terrain and strong winds and that many have developed life cycles that are carefully synchronized with the pattern of winter rain and summer drought characteristic of the south-western Cape.

In summer, the permanent high-pressure system in the atmosphere over the south Atlantic Ocean moves further south to a position west of the Cape Peninsula and diverts the rain-bearing frontal systems that travel east across the Atlantic, preventing them from passing over the Cape. At the same time, the high pressure cell is responsible for the south-easterly gales that batter the Peninsula in summer. These south-east winds blow anti-clockwise around the high-pressure cell and are forced out over the warm waters of False Bay from where they pick up moisture. When the south-easter blows upwards over the mountains, including Table Mountain, it cools and forms a thick cloud of condensed vapour – the famous tablecloth that falls over the northern slopes of Table Mountain and keeps the summit moist even when the lower slopes are parched and dry.

In winter, the Atlantic high pressure cell moves to a slightly more northerly position. Winter weather is dominated by a succession of cold fronts that strike the west coast as rain-bearing north westerlies. The amount of rainfall received at different points on the Peninsula varies, depending on position in relation to the mountains. Rainfall (and moisture deposits from

Thunder and lightning at Scarborough, Atlantic coast.

cloud and mist) measures only 350 mm at Camps Bay (on the west side of the mountain) and as much as 2 000 mm at Newlands on the south-east slopes.

In these extreme and variable conditions, plants (in collaboration with insects and small rodents) have developed remarkable strategies for survival. Many fynbos species are uniquely adapted to withstand fire – an inevitable occurrence in the hot dry Cape summers – and some even depend on fire for

germination and seed dispersal. In some instances, ants gather seeds with nutritious coatings into underground chambers where, safe from predators, the seeds eventually germinate. Under natural conditions, fires are likely to start as the result of lightning strikes. In recent times, however, the too-frequent burns – almost always started by humans – and the intensity of the fires fed by alien trees species, are making it more difficult for fynbos areas to regenerate.

The three plant species that characterize fynbos vegetation – proteas, ericas and restios – have all evolved defence systems against the hot windy summers. Some proteas have large leaves covered with tiny hairs that reflect sunlight; the masses of tiny, slightly-curled leaves of ericas present a considerably reduced surface area, thereby reducing evaporation; and restios, or reeds, have dispensed entirely with visible leaves.

In contrast to the Peninsula's diverse and adaptable plant species, the large mammals once evident in the region – elephant, black rhino, eland, the Cape lion and the Quagga (both now extinct) – have virtually disappeared. The last lion was killed on Table Mountain in 1802, eland disappeared in 1840 (although subsequently re-introduced), and the leopard appears, from evidence at Cape Point, to have survived until relatively recent times.

There is still much to be seen of wildlife, but it takes interest, time, patience and some luck. One of the objectives of this book is to give a glimpse of timid and seldom-seen creatures: Cape grysbok, Cape porcupine, genet, water mongoose and Cape fox. Snakes – among the larger predators in the area – are visible in most of the habitats shown here and still make an appearance in Peninsula gardens.

Visitors to the natural areas of the Peninsula are likely to see more birds than any other vertebrates, although few species are endemic to fynbos. Among these, the Cape sugarbird is, possibly, the most conspicuous, gathering in numbers in winter to feed on

Rock art in Peers Cave above Fish Hoek.

nectar from protea flowerheads and to pollinate them in the process. Like the myriad insects and reptiles and amphibians, including snakes, lizards, tortoises, frogs and toads that inhabit the region, birds are an intrinsic part of the intricate relationships that work to the benefit of each and all.

It is difficult to establish when humans first entered the south-west Cape, but there is evidence to suggest that the Peninsula was inhabited by early Stone Age man close to a million years ago. During the last of the so-called ice ages, twenty thousand years ago, the sea level was about 140 metres lower than at present and early humans would have explored and exploited plains, valleys and hills that today lie beneath the waves of the Atlantic. In more recent times, the Peninsula was home to strandlopers and herders whose way of life was changed forever in the wake of European exploration and settlement.

Although humans appear to have lived for many thousands of years within the natural flux and flow of their environment, the advent of commerce and technology caused drastic alterations. Indigenous vegetation was cleared to make way for agriculture and ancient forests growing on the higher mountain slopes were converted into wagons, ship's planking, furniture, buildings and firewood. Large mammals were driven out of the area or hunted to extinction. In the 1890s, planting of alien forests began.

Humans may be conspicuous by their absence in the pages that follow. The urban sprawl, historic buildings, cultural artefacts, commercial lures, even conservation efforts that have helped to eradicate alien plant species and to redeem human presence in the area, are not here. This is a book about an environment that is ultimately dependent on the willingness of ordinary people to conserve it. But it is mostly about an environment that has managed to retain its complex and beautiful aspect inspite of us.

Fossilized tooth of a great white shark.

Several different habitats coexist in this 'continental cul-de-sac':

OCEAN AND SHORE

The coastline stretches from Table Bay, on the Atlantic Ocean side, south and east to the waters of False Bay.

WESTERN STRANDVELD

A narrow belt of shrubland on the seaboard's ancient marine beds (not to be confused with 'sandveld', a geographical region).

STREAMS, LAKES AND WETLANDS

Inland water bodies, often fringed with dense reedbelts and fed by streams from mountain catchment areas, rain and mist.

LOWLAND FYNBOS

Once extensive on the sandy soils of the Cape Flats, but now limited to a few small enclaves, two of them in popular racecourses.

RENOSTERVELD

Related to fynbos, renosterveld grows on more fertile soils on Signal Hill and at the foot of Devil's Peak and Lion's Head and at Tygerberg.

MOUNTAIN FYNBOS

The dominant vegetation type, which covers all the mountains in the area and contains a great diversity of plant species.

AFROMONTANE FOREST

Temperate afromontane forest grows in gorges and ravines on the eastern and southern slopes of Table Mountain.

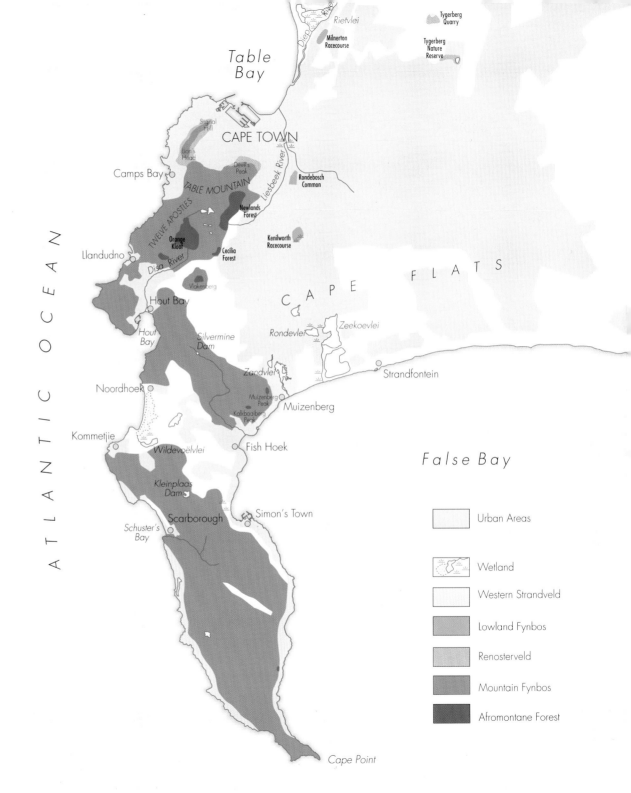

ATLANTIC OCEAN

Table Bay

CAPE TOWN

Camps Bay

Llandudno

Hout Bay

Hout Bay

Noordhoek

Kommetjie

Schuster's Bay

Scarborough

Simon's Town

CAPE FLATS

Strandfontein

Muizenberg

Fish Hoek

False Bay

Cape Point

Rietvlei

Milnerton Racecourse

Tygerberg Quarry

Tygerberg Nature Reserve

Signal Hill

Lion's Head

Devil's Peak

TABLE MOUNTAIN

Liesbeek River

Rondebosch Common

Newlands Forest

TWELVE APOSTLES

Orange Kloof

Cecilia Forest

Kenilworth Racecourse

Disa River

Vlakenberg

Silvermine Dam

Rondevlei

Zeekoevlei

Zandvlei

Muizenberg Peak

Kalkbaaiberg Peak

Wildevoëlvlei

Kleinplaas Dam

Legend

- Urban Areas
- Wetland
- Western Strandveld
- Lowland Fynbos
- Renosterveld
- Mountain Fynbos
- Afromontane Forest

OCEAN AND SHORE

the continent's edge

The Peninsula's coastline extends from Table Bay south to Cape Point and east along False Bay. The shoreline is primarily rocky, with steep and dramatic cliffs, but there are long stretches of sandy beach, many of them littered with granite and sandstone boulders that tumbled down the mountain slopes millions of years ago. Two mighty currents meet nearby at Cape Agulhas: the warm Mozambique-Agulhas Current and the chilly nutrient-rich Benguela Current that sweeps up the west coast from its source among the icebergs of the Antarctic. This plankton-rich flow from the Antarctic is the beginning of a food chain that attracts fish, sea mammals and birds in large numbers to the coast. Dense kelp forests are a prominent feature of more shallow waters at the shore and, in the warmer waters of False Bay, tidal pools shelter many distinctive forms of marine life.

14

1

A flock of common terns (1) scatters upwards at Olifantsbos on the Atlantic coast in a flurry of white underparts — to settle some distance away from the intruder. These highly gregarious birds are visitors, arriving in early summer after a long journey from their northern-hemisphere breeding grounds, to winter in the Cape. Their days are spent roosting on boulders or flying swallow-like above the water searching for small fish. Also seen off the southwest coast, the Black-browed Albatross (2 & 4) soars in effortless flight, gliding for hours in search of food. It touches ground only when it comes down to breed – on a number of small and remote islands in the southern ocean. The Yellow-nosed Albatross (3) is less common and is smaller, but no less impressive.

1

In the more temperate waters of False Bay, a southern right whale and her calf (1) move past the rocky shore. These huge mammals arrive year after year in midwinter to mate and give birth to a single calf. By December, however, they will have negotiated the waters off the Cape, heading for their feeding grounds near Antarctica, 5 000 km away. All around the coast lie the bulky wrecks of other leviathans that have been less successful in weathering the Cape's stormy seas. The American vessel, *Thomas Tucker* (2) – now a breeding site for cormorants – hit ground in 1942 near Olifant's Bay on the Atlantic coast, with a cargo of army tanks.

1

There has been a dramatic increase in the number of Cape fur seals around southern Africa over the last few decades; their status has changed from 'Endangered' to 'Common'. Like African Penguins, these blubbery oceanic predators favour off-shore islands as breeding sites, including the famous Robben Island (the name meaning 'seal' in Dutch). This colony (1), numbering a few thousand, made its home on Duiker Island, off Hout Bay. Comfortable on the rocks, a young seal cub (2) snoozes sweetly, forgetting the world around it. Another youngster, an elephant seal cub (3) — rarely seen in these parts — rests after an exhausting journey from the southern ocean to Olifant's Bay where it has come in to moult.

The African Penguin (formerly known as the Jackass Penguin) is the one penguin species that breeds exclusively off the south-western African coastline, mostly on islands. However, in 1982, two of these birds (whose total numbers have been decimated, largely by man, during the last hundred years) were discovered nesting on the mainland about 20 km from Cape Town, at Boulders Beach. They had chosen to inhabit suburbia and a public beach and, despite a traumatic history of dog and cat attacks, poaching and irate householders, have increased in number from 2 to more than 2 500 penguins.

The cliffs at Cape Point (1) rise dramatically from the Atlantic Ocean and are considered to be the highest in the world. Composed of sturdy, reddish Table Mountain sandstone and weathered by wind and water over millions of years, their nooks and crannies offer shelter and breeding sites to hundreds of sea birds. The lighthouse with its spotlight (the most powerful on the African coast) warns of hazards in these treacherous waters, where many mariners have found a watery grave. Further north, at Smitswinkel Bay, sandstone rocks (2) litter the otherwise sandy beach. The erosion of rocks such as these, originally from higher-lying areas, has produced much of the Peninsula's sandy soils. When exposed, the rocky surfaces are soon colonized by lichens and mosses, creating interesting patterns in colours that contrast with the yellow-red surface of the sandstone.

Tidal pools, teeming with life, are a highly specialized feature of the oceanic ecology of False Bay, and
exploring them is a fascinating pursuit. Shallow rock
basins, hollowed out over millions of years, are filled
afresh by the tide with water rich in food for the
myriad creatures that inhabit these small and enclosed
but stressful refuges. Cape urchins (1 & 3) in red, pink
and purple, with razor-sharp quills, are abundant,
as are plate-sized starfish that prey on mussels and
other shellfish. The colours and patterns of dwarf
cushionstars (2) vary greatly. Extremely well camou-
flaged, they are often hard to see with the naked eye.
Marine life in the four inter-tidal zones of the rocky
shore (4) varies with degree of exposure and the force
of wave action.

2

The waters of False Bay (1) merge with the sky in a soft pink-purple haze shortly after sunset, the last whispers of the dying day. Seen at a distance, Table Mountain (2) does not reveal the busy city life at its feet. The flat (1 080 metre high) table top is flanked by the freer-flowing curves of Devil's Peak on the left, and Lion's Head and Signal Hill on the right.

WESTERN STRANDVELD
built on sand

Along the coast of the Peninsula, on the alkaline sands of ancient marine beds, grows a coastal vegetation known as strandveld. Among the old and long-lived bushes and shrubs that form climax strandveld are Cape sumach and sand olive trees and — in places like Cape Point that are protected from urban development — thickets of milkwood, well adapted to the coastal winds. Sustained by the slightly more nutrient-rich soil, containing tiny fragments of calcium-rich seashells, many of the strandveld shrubs are able to produce berries and fruit, which support small mammals and birds as well as reptiles. Strandveld is not as dense as fynbos vegetation and is not dependent on fire for regeneration, but on its many small inhabitants. The second type of strandveld, pioneer strandveld, is well preserved at Rondevlei, Zandvlei and around Cape Point. It is distinguished by the bietou bush and other small bushes as well as by carpets of fleshy-leaved sour figs and by flowering annuals that reach their glorious peak in spring.

1

Footprints in the sand. The fresh spoor **(1)** of a Cape clawless otter along a dry sandy riverbed is evidence of the presence of this fierce seafood hunter. Other clues are 'latrines' of droppings made up largely of crushed crab shells. Because of their nocturnal life-style, otters seem to be able to coexist successfully with humans. On rare occasions, a pair or small family group may be seen hopping nimbly over rocks in the inter-tidal zone **(2)** – an unusual sight and a special thrill for the lucky observer.

2

At high tide, brackish sea water, filtered through tonnes of coastal sand, fills a shallow depression (1) beyond overgrown sand dunes on the Atlantic coast. In a close-up view, the open spreads of ivory-coloured dunes rippled by the wind (2) present a desert landscape, softened only a little by clumps of pioneer strandveld in the distance.

1

2

The Amoured Ground Cricket (1) and the Common Dotted Border (2) are just two of the many insect species that inhabit the strandveld. Butterflies are among the insect prey of various lizard species, most of which are active by day and favour rocky or open areas. Knox's Desert Lizard (3) is a small agile reptile that may occasionally be seen rushing along sparsely vegetated sand in search of a safe place to shelter.

3

1

Minaret flowers (1) on a plant known locally as wild dagga and scientifically as *Leonotis leonurus*, bloom for the Lesser Double-collared Sunbirds that visit them in numbers in spring. *Satyrium carneum* (2) the largest orchid in this habitat, makes a quiet display in spring-time garb, while the bietou bush *Chrysanthemoides monilifera* (3) flowers in careless abundance. It is seen here in early summer, as are the Long-Horn Beetles mating on a *Ruschia macowanii* (4) at Cape Point. (This genus, *Ruschia*, does not have the fleshy fruit that generated the family's common name 'vygie' or fig.) *Hyobanche sanguinea* (5) is a fleshy, sticky parasite that lives on (and off) the roots of other plants in the strandveld.

Although fairly common, the porcupine (1) and large-spotted genet (2) are nocturnal creatures and are not often seen. Plants that have been dug up for their tubers or fleshy roots and twigs stripped entirely of bark or fruit are evidence of the presence of the porcupine – Africa's largest rodent. Large-spotted genets are most often seen in branches of trees or running across the road. The grysbok is a fynbos endemic that tends to remain in more densely vegetated areas. This newborn fawn (3) is still finding its feet.

Unlike dunes that the wind shifts year in and year out, a few centimetres at a time, these overgrown dunes at Cape Point have been stabilized by vegetation. A very different vista emerges in unprotected areas, where the strandveld has succumbed to development.

Further north, at Rondevlei, dew clings to bushes and restios on a chilly winter morning. Although rain falls regularly here in winter, mist and cloud are an important source of moisture. Restios are particularly adept at catching this ethereal water.

STREAMS, LAKES AND WETLANDS

freshwater oases

A number of streams – some perennial, others flowing only in winter – run down the mountain slopes of the Peninsula towards the sea. They are fed by rain, but also by mist and cloud that condenses on the mountain tops. In turn, some feed rivers and inland lakes on the flatlands that offer prime habitat for a variety of plants and animals. Although the mountain streams are pristine, the lower reaches are mostly canalized and degraded by urban pollution, with eroded banks and infestations of alien vegetation. In a desert of urban development, inland lakes such as Rondevlei, Zeekoevlei and Strandfontein are oases of peace and tranquillity. Their extensive adjoining reedbelts provide shelter for thousands of birds to roost and breed. Equally important to plant and animal life are numerous temporary wetlands – pans like Rietvlei and pools at Kenilworth Racecourse that are filled with water in winter and dry out in summer.

44

1

Greater Flamingos (1) are common summer visitors to the Peninsula, where lakes rich in micro-organic life provide energy-rich food for the long journey back north. Flocks ranging in number from a few hundred to over a thousand gather at large bodies of water, as at Zeekoevlei. Spectacular in flight, flamingos need some fancy footwork (2) on the surface of the water to gain sufficient acceleration to take off.

Groups of White Pelicans, massive birds with huge bills, long necks and short legs, are best seen at Zeekoevlei, Rondevlei and at Strandfontein, where their numbers fluctuate as they are not a local breeding species. Side by side, in flocks of more than a hundred, they slowly make their way through the water, filtering out small to medium sized fish, which they swallow whole. Once airborne, they are excellent gliders with a graceful flying style, their wings spanning a length of more than three metres.

1

Visible singly or in small groups, the all-white African Spoonbill (1) may nest with other species, generally in reedbeds or dead trees. The White-breasted Cormorant (2) is the largest of the five South African cormorant species. Like the endangered Black Oystercatcher (3), it is found in both freshwater and marine environments. In a shallow stream at Rondevlei, a Little Egret (4) stands motionless, looking for prey.

1

Surrounded by the growing urban sprawl on the Cape Flats, the Peninsula's lakes and adjacent wetlands are a relaxing and rewarding refuge for humans as well as plants and animals. Although the water level in some areas falls in summer – considerably so in dry years – enough is left for inhabitants like this hippo bull **(1)**, caught here in a fearsome display of dominance over family members. At Rondevlei **(2)**, over 225 species of birds have been recorded since the reserve was proclaimed in 1952.

2

Fish eagles appear to be unperturbed by human encroachment. There are resident pairs at Zeekoevlei and Zandvlei and even a couple of breeding pairs at Rietvlei and the Cape of Good Hope section of the Peninsula National Park. Here, a sub-adult female at Zandvlei tries to keep her fishy meal out of the reach of her mate. These huge raptors mate for life and usually keep to a designated area – a lake or section of a river. Unmistakable with its white head and breast, the fish eagle's haunting call is one of the sounds characteristic of the African wilderness.

In golden light, the reeds are reflected off the water in a shallow pan near Zandvlei just before sunset **(1)**. On the other side of the pan, a Grey Heron **(2)**, generally a still bird that spends its time waiting for prey, takes off from its roosting site. The quest for food, mostly freshwater crabs, brings the water mongoose **(3)** into the open. Small animals and birds find good cover in the dense reedbelt bordering the natural channel **(4)** that connects the lagoon at Zandvlei with smaller ponds and pans.

56

4

1

Early morning in summer and the distant city is yet to come alive (1). A Purple Gallinule (2), with its distinctive frontal shield and red bill, hurries towards shelter in the reeds, while a Black-winged Stilt (3) steps across the mud, picking insects, snails and worms off the surface. In flight, the long red legs of this slender black-and-white wader trail conspicuously behind it.

1

Three species of kingfisher are resident on the Cape Peninsula. Here, distinctive black-and-white Pied Kingfishers **(1)** perch on fallen wood. They are often to be seen hovering above the surface for quite a while before swooping down to spear a fish with their razor-sharp bills. Their cousin, the Malachite Kingfisher **(2)** – smaller in reality than they are, though not in the perspective of this picture – is dressed in rainbow colours of shiny red, orange and deep blue. It prefers to scan its hunting grounds from an exposed branch or reed stem.

2

Pristine water. On bank vegetation near a Table Mountain stream, a shimmering damselfly (1) is poised with balletic grace. These insects depend on freshwater streams for breeding grounds into which to drop their eggs. Elsewhere, a Leopard Toad (2) – a rare species that may be encountered crossing the road at night – heads towards its own breeding grounds in early spring. Between the fine leaves of a reed, pearls of pure water hang on the threads of a spider's web (3) on a winter morning.

There are 21 species of frog to be found on the Peninsula, one of which (the endangered Table Mountain ghost frog) is endemic. The Cape River Frog (1), common in streams on Table Mountain, is to be found even in very small ponds, always on the look out for a food item. It is more often heard than seen — escaping with a splash into the water. The Cape Dwarf Chameleon (2) is fairly common and seldom strays far from water, but is difficult to detect because of its mercurial ability to change colour. Also common, but no less lovely for that, are these water lilies *Nymphaea nouchali* (3) on a pond at Silvermine.

1

2

3

1

Long dry summers mean that many plant species must make special adaptations to survive. The leaves of an ice plant *Mesembryanthemum crystallinum* (1) consist of a fleshy centre wrapped in a sturdy skin to store moisture effectively for long periods. At the large pan at Rietvlei, home to thousands of waterbirds and waders, the water dries up completely during summer, leaving behind a mudplain with a tightened and cracked skin (2).

2

LOWLAND FYNBOS
refuge for rarities

A few patches of lowland fynbos survive on the Peninsula, characterized by restios and ericas, two of the main components of fynbos. Proteas are less common than in mountain fynbos, and trees are not often seen, but bulbous plants or geophytes are numerous and burst into flower after good rains in early spring. This is a fire-driven system that needs to be burnt at regular intervals (ideally every 20 years or so) for shoots to sprout and for some plants to bloom in a temporarily competitor-free environment. The most important preserve of lowland fynbos on the Peninsula is the centre of Kenilworth Racecourse, one of the last bastions of a number of endemic lowland fynbos species and host to the endangered micro frog, the rarest frog species in Africa. Rondebosch Common survives amid surburban housing and development — a reminder of the landscape of a hundred years ago. Milnerton Racecourse and patches of sandplain fynbos on the Cape Flats are the other remains of a seemingly expressionless habitat that holds an intricate weave of plant and animal life.

A small tract of land at Rondevlei (1) contains a number of extremely rare species including the yellow-flowering *Serruria aemula* and (in the foreground) *Erica margaritacea*, a site-endemic that grows nowhere else in the world. Tracts of sandy lowland fynbos, such as the centre of Milnerton Racecourse (2) support the summer-flowering *Orphium frutescens* and other plants including pelargoniums (3) – species of which have parented the geraniums seen in European window boxes. The Parrot-beaked Tortoise (4) is one of the less common inhabitants of lowland fynbos, but is still to be seen at Kenilworth Racecourse, the least-disturbed stretch of lowland fynbos on the Peninsula. The beautiful white-bellied Aurora House Snake (5) is, increasingly, a rare sight.

The striped field mouse **(1)** can be found in many different habitats throughout the fynbos biome. It eats seeds, playing a vital role in the ecosystem by controlling thus the number of seedlings that can grow. It also helps to disperse seeds that it digests only partially as well as those it scatters on its journeys to its larder. The large Cape dune molerat **(2)**, which can attain a weight of three-quarters of a kilo, also plays an important part in dispersal of the bulbs, corms and rhizomes on which it feeds, ploughing the ground into long furrows and throwing up mounds of sand in the course of its extensive excavations. Similar in appearance, although much smaller and with pretty facial markings, the Cape molerat **(3)** leads a life much like that of its large brown cousin.

1

2

In typical pose, a Praying Mantid (a species known locally as the Hottentot god) **(1)** awaits its insect prey in a garden near the Rondebosch Common. A velvety Rain Spider **(2)** on the Common crafts a solid nest of web-wrapped plant matter in which its eggs will be laid and its young safely nurtured. Still closed, the flowers of *Baeometra uniflora* **(3)** bloom alongside a dew-dotted 'vygie' **(4)**, a member of the prolific Cape family of mesembryanthemums.

A well-known sound – to be heard even at the dead of night in central Cape Town – is the call of the Crowned Plover **(1)** which breeds in low-lying clumps of vegetation in sandy flatlands, where it zealously guards its nest. In the same habitat, an African Monarch **(2)** seeks its supper on a fynbos blossom. Butterflies are not common in fynbos, possibly because the leathery and nutrient-poor vegetation does not provide enough food for their larvae. Not many proteas occur in lowland fynbos; *Leucadendron levisanus* **(3)** is an extremely rare member of the family, once thought to be extinct.

3

The centre of Kenilworth Racecourse **(1)** retains much of its fynbos wealth and affords a glimpse of what the Cape Flats must once have looked like. On the sandy soil, *Orphium frutescens* **(2 & 4)** blooms in profusion, often in marshy areas. *Ixia monadelpha* **(3)**, another typical member of the lowland fynbos community, though not as common, emerges phoenix-like from a bunch of restios. Stringent conservation measures are needed if these and other species are not to be lost.

RENOSTERVELD

before vineyards and fields

Renosterveld is named for the grey-looking renosterbos, meaning rhinoceros bush, the most prevalent shrub in the area. The rhinoceros connection is not quite clear, although rhinos were among the grazing mammals that roamed the Cape at a time when renosterveld was widespread. Today, more than 95 per cent of renosterveld has fallen under the plough and, within the area of the Peninsula, only a few isolated pockets survive, on the lower mountain slopes. The relatively fertile clay-rich renosterveld soils make it more suitable for farming than the sandy acidic soil of fynbos areas. It supports more nutritious vegetation than fynbos and a variety of grasses. In fact, renosterveld may have resembled grassland before it was conquered by renosterbos. Proteas are rare and restios almost non-existent, but a variety of bulbs flower in season, in spectacular displays of colour and vibrance.

Drab and uniform as the small-leaved shrubs of the renosterveld may look at a distance, closer investigation reveals a variety of plants and, after good rainfall, numerous species of ground orchids and an abundance of bulbous annuals. They include *Ixia odorata* (1), an eye-catching member of the iris family. Another yellow flowering annual is the autumn star *Empodium plicatum* (2), which blooms vividly with the onset of the winter rains. Two years after a fire, young growth can be seen on Signal Hill (3). Proteas such as the *Leucadendron salignum* in the foreground are unusual in renosterveld.

3

Rhinoceros bush (*Elytropappus rhinocerotis*) on Signal Hill sports tiny pale-green shoots in late autumn **(1)**, a screen for the *Leucadendron salignum* in the background. At ground level, the male flowers of *Arctopus echinatus* **(2)** spread over the light-brown clay soil. *Euphorbia caput-medusae* **(3)**, another low-lying rambler, forms round colonies up to a metre in diameter on the ground.

The Cape Rain Frog (1) emerges only after the first rains at the end of summer, spending the dry months completely buried in soil. Although present in many habitats, the Brown House Snake (2) is rare in the Peninsula. (This picture may be a unique record.) The Cape Cobra (3), another of over 20 species of snake on the Peninsula (seen here in its juvenile stage) is a shy and nervous snake that spreads its hood to warn off trespassers in the dry bush. Renosterveld does not receive valuable moisture from fog and mist (4).

1

Insects occur in abundance in renosterveld, using the variety of shrubs for food and shelter. The bright yellow and red of this Foam Grasshopper *Dictyophorus spumans* (1) warns of the unpleasant foam it produces when attacked – the kind of surprise it is wise to expect from brightly coloured and boldly patterned insects. The antlion (Genus *Palpares*) (2) is large and showy in its adult form, but emerges from a surprisingly short and stout pupa and an ugly predacious larva, which buries itself in small conical pits of its own making – a familiar sight in sandy soil.

2

On the lower slopes of Devil's Peak and Signal Hill, clay soils nurture great numbers of annuals. Among them are the March lily *Amaryllis belladonna* (1), which blooms in late summer, *Oxalis purpurea* (2), and an *Erepsia* flower (3), struggling for a glimpse of sunlight from under dense shrubs. After a field fire, the spring rains bring out a magnificent display of thousands of *Watsonia borbonica* (4).

4

MOUNTAIN FYNBOS
amazing diversity

The largest area of the Cape Peninsula, by far, is covered with mountain fynbos. Despite the invasion of alien plants, the indigenous species flourish (under the auspices, now, of the Cape Peninsula National Park) on the slopes and summits of the Peninsula's mountains. Within the fynbos biome, mountain fynbos contains the highest number of plant species, a large number of them to be found on Table Mountain. The four plant types that characterize fynbos – proteas, ericas (heaths), restios and geophytes (plants with bulbs and corms) – thrive on the leached and acidic sandy soils, their numbers and growth dependent on altitude, aspect, season, wind and fire. There are few indigenous trees (although some proteas such as the silver tree can grow to almost 12 metres tall), and not many large mammals, but the vegetation supports a great diversity of insect life and many small creatures. Three of the six bird species endemic to fynbos may be seen in mountain fynbos on the Peninsula – the Cape Sugarbird, the Orange-breasted Sunbird and the Cape Siskin.

Pristine fynbos vegetation (1) crawls up the lower mountain slopes on the Atlantic seaboard, which rise steeply to form a narrow mountain chain contiguous with Table Mountain – the Twelve Apostles. (In fact, there are more than 12!) Although it seems dull and unattractive at times, fynbos contains a host of seasonal flowers and is home to many small mammals. The small grey mongoose (2) may be seen regularly raiding litter bins in search of tasty morsels. Remarkably resourceful when it comes to looking for food, the chacma baboon (3) is another scavenger at garbage bins, regularly seen at Cape Point. The Cape fox (4), a beautiful, dainty and strictly nocturnal animal – one of the few carnivores found exclusively in southern Africa – is a rare sight.

The Orange-breasted Sunbird (1), endemic to the south-western Cape, gathers in numbers on flowering ericas to probe for nectar and pollen with its specially adapted beak. The Cape Sugarbird (2) is easily identified by its long tail (particularly in the male) and is the species most people associate with fynbos because of its noisy presence in protea shrublands in autumn and winter. Another fynbos inhabitant, the small Grassbird (3), hides most of the time, but climbs on exposed twigs in the early morning to fill the air with its melodious call at places such as this erica meadow (4) at Cape Point.

With the changing seasons, flowers appear in the fynbos in all their natural diversity. The *Wachendorfia brachyandra* (1), seen here fringed with restios at Cape Point, prefers damp ground, After good winter rainfall, *Spiloxene capensis* (2) and other annuals form flower carpets on the slopes of Lion's Head (3). *Dilatris corymbosa* (4) are often seen at Silvermine in springtime and early summer, while a sturdy *Brunia* (5) blooms near Table Mountain's lower cable station, attended by a small metallic-green fly. *Cotyledon orbiculata* (6) occur in large numbers in many areas, often flowering in profusion in midsummer. They are a familiar sight on the Peninsula.

There are 25 species of snakes on the Peninsula, 6 of them venomous, 4 of them potentially deadly. Few are restricted to the fynbos community of plants and animals, however, and those that do live there are seldom encountered. Like most snakes, the rare Berg Adder **(1)** is diurnal. Short and sluggish, though sturdily built, it is equipped with a powerful venom that immobilizes its vertebrate prey instantly. Like the more often encountered Southern Rock Agama **(2)** – a miniature dragon, with spiny scales around its neck – it is fond of basking in the sun on rocky ledges.

1

2

Large carpets of mesembryanthemums **(1)** add colour to the drab green and beige vegetation. The flowers may look dried out, but the thick fleshy leaves that sprawl on the ground beneath them are a valuable storehouse of moisture. At higher altitudes, on Table Mountain and elsewhere, *Aristea macrocarpa* **(2)** blooms in spring. An African Hummingbird Moth *Macroglossum trochilus* **(3)**, one of the myriad insects attracted to fynbos, remains airborne while it sheds the petals of its pollen-provider with the rotor-like movement of its wings.

2

3

1

2

Unlike other parts of southern Africa, the Peninsula does not afford much game viewing. Nowadays, bontebok (1) may be seen on undisturbed coastal fynbos at the Cape Point section of the Cape Peninsula National Park. Almost extinct in the early 1930s, bontebok numbers have steadily increased through conservation efforts, but they are still a rare species. Eland (2), the largest African antelope (old bulls weighing as much as 800 kg) have been introduced in flat areas at Cape Point.

A small herd of Cape mountain zebra **(1)** roams the plains and lower slopes at Cape Point. Never numerous, in all likelihood this inhabitant of mountain ranges in the Cape was relentlessly hunted and by the mid-1930s was almost extinct. Conservation efforts have led to a remarkable recovery in numbers, primarily in the Eastern Cape. With its low nutrient content, fynbos cannot support large numbers of bulk-feeders like zebra. This elevated plateau-like plain **(2)** at Cape Point, with its soft hill slopes beneath steep rocky peaks overlooking False Bay, is large enough to enable small herds to survive.

1

2

Conspicuously bright and shiny, *Erica cerinthoides* (1) is pollinated exclusively by the beautiful Mountain Pride butterfly (Table Mountain Beauty), which is strongly attracted to red flowers and is the exclusive pollinator of many plant species. Mountain dahlias *Liparia splendens* (2) and *Agapanthus africanus* (3), inhabit high-lying areas of the Silvermine reserve. *Salvia chamelaeagnea* (4) bloom above Chapman's Peak Drive, back-lit by the setting sun.

3

4

Disas, the prettiest of the orchids, inhabit the Silvermine section of the Peninsula National Park (1) among other areas. *Disa racemosa* (2) requires moisture to prosper, but *Disa cornuta* (3) grows in lowland sandy plains and on mountain peaks. The blue disa *Disa graminifolia* (4), seen here on top of Table Mountain, forms small, but colourful flower heads in late summer. *Disa atricapilla* (5) grows at Silvermine, and the famous red disa *Disa uniflora* (6) flourishes in moist areas, usually along mountain streams.

Like a spirit, a Green Mantid **(1)** stands out against the sun on the back of a back-lit leaf. Often appearing in gardens in the Peninsula, these ferocious hunters are very effective agents of natural insect control. Elsewhere, a beetle probes the head of a pincushion protea **(2)** for pollen. Bright yellow *Disa tenuifolia* **(3)** bloom at a moist fynbos patch at Silvermine, together with other delicate orchid species, in late spring.

3

1

More than 40 species of protea inhabit the Cape Peninsula, many of them on Table Mountain. Well-named for its studded head, the pincushion protea – of which there are several species – forms dense stands (1) on the back table of the mountain. When open, the large king protea (2), has a flower head about the size of a dinner plate. Another protea, the wagon tree (3), common on the Peninsula, was so named by Dutch settlers who used the sturdy wood of its trunk for the wheel spokes of their wagons. The smallest protea of them all, *Diastella divaricata* (4), has flowers the size of a thumb nail.

More proteas. The iridescent leaves **(1)** of the silver tree are covered with fine silvery hairs that reflect sunlight and protect the plant against excessive exposure to the sun. Silver trees are at their most attractive in summer in a strong wind that makes the leaves shimmer. They are endemic to the Cape Peninsula and the seeds in their woody cones are released by fire. They are best seen at Lion's Head **(2)** and the higher-lying areas of the Kirstenbosch Botanical Garden. Silver trees are not long-lived and it is said that a silver tree will flourish only within sight of Table mountain – a myth that very likely stems from its propensity to sudden death from fungal infection.

Moist fynbos areas, usually along streams or around ponds and flooded depressions, support distinct, highly specialized plant communities of restios, grasses, ferns and a great many delicate orchid species. Along a small perennial stream on Table Mountain, fringed by ferns, the bright red petals of the red disa (1) stand out like sparks. *Drosera hilaris* (2), an insectivorous sundew, traps and 'eats' small insects held by sticky glue-like drops in its 'fangs'. *Pterygodium acutifolium* (3), a beautiful ground orchid, flourishes in damp soil, shaded and protected by dense stands of restios.

Like the rock itself, a well armoured Cape Girdled Lizard **(1)** stands motionless on a mountain outcrop, remnant of millions of years of erosion. Nearby lies a coiled Puff Adder **(2)**, a sluggish creature but poisonous and ready to bite when cornered or inadvertently trodden on. Rock dassies **(3)**, a Peninsula favourite, warm themselves on the rocks in the morning sun. Their appearance notwithstanding, they are not related to rodents, their nearest relatives, (in taxonomical terms) being elephants! In the quiet tranquillity of early morning, the inhabitants of Cape Point go about their business, barely perceptible in the sandy fynbos stretching to the Atlantic shoreline **(4)**.

4

AFROMONTANE FOREST

in dark places

Evergreen indigenous afromontane forest — not to be confused with tracts of cultivated pine and eucalyptus plantations — grows in ravines and gorges on the wet southern and eastern slopes of Table Mountain. Extensively exploited by early settlers for their valuable timber, many old pioneer species such as yellowwood, Cape beech, stinkwood, wild peach and saffronwood survive in relatively poor soils, enriched by a layer of humus. They grow along the Disa Stream at Orange Kloof and further along the southern face of the mountain. At higher altitudes, tree branches are wound round and round with parasitical growth and bushes, mingled with lianas from above and a mat of mosses and lichens below, make the forest almost impenetrable. Crevices and caves, their entrances hidden by trees, are a feature of the forest. A number of typical birds such as the Rameron Pigeon, Cape Batis and Forest Canary may appear throughout the year; others, like the Paradise Flycatcher, are summer visitors.

Colonized by moss and ferns (1), a vertical wall of rock stands under the forest canopy, dry except for a few trickles of water. In winter, a raging torrent will tumble over it, hiding the lush, green vegetation beneath. Nearby, a tree trunk (or two?) twists upwards towards the light (2), and a Red-breasted Sparrowhawk (3), a small but agile hunter, makes its way through foliage. It is mostly seen in flight, in search of the small birds that are its prey. In autumn, April Fool flowers *Haemanthus coccineus* (4) bloom in their hundreds on the Spes Bona forest floor, on the slopes above Kalk Bay.

1

On the back table of the mountain, forest growth covers the entrances to a number of caves – deep, dark caverns, cool and humid, that serve hundreds of Egyptian fruit bats (1) as daylight roosting spots and hibernation sites. Even the smallest cranny (2) provides these winged miniature foxes with a grip. When dusk approaches, a wave of unrest flows over the colony and, shortly thereafter, its inhabitants flutter out into the night, one by one, in search of food. By far the largest bats on the Cape Peninsula, these big-eyed creatures may be detected by the mess of discarded fruit and droppings that accumulates underneath their feeding perches.

2

At higher altitudes, *Usnea lichens* or Old Man's Beard grows over the ancient forest canopy **(1)**, waving gently in the wind. Water is the life blood of indigenous forests, which require at least 500 mm of rain a year to grow, particularly in winter rainfall regions. In summer, waterfalls on mountain faces **(2)**, as at Silvermine, are reduced to a mere trickle. However, the moisture-bearing southeasters that blow over False Bay and are pushed up the mountain side help to sustain dense foliage and, where the sun's rays touch the forest floor **(3)**, dozens of herb species flourish. Frightening, possibly even nightmarish, to the uninitiated, the forestscape nevertheless has many charms.

3

The largest remnant of indigenous forest on the Peninsula is found at Orange Kloof **(1)** along the Disa stream, on the back table of the mountain. On the fringes of the forest, the larvae of fireflies hatch **(2)**, their fully developed shiny tails already betraying their bright presence. The males that hatch can fly but females remain wingless 'glow worms'. High above them, a forest giant **(3)** reaches for light and, in the process, spreads darkness on everything beneath its leafy crown. Paper wasps **(4)**, another of the insect species that make their home in the forest and in other habitats, make large nests on the underside of rocky overhangs. In winter, the female wasps will abandon this elaborate 'paper' configuration and hibernate in any sheltered place.

Small and easily overlooked, a Marbled Leaf-toed Gecko **(1)** creeps over a piece of bark on the forest floor. One of two gecko species on the Peninsula, this flat little nocturnal creature enters houses at night or hangs around outdoor lights awaiting its insect prey. On a rocky outpost, a wizened weed provides its own reflected light **(2)** while a curious Spotted Eagle Owl chick **(3)** stares unblinking at the camera from the safety of a tree. These raptors are sometimes seen at night in and above central Cape Town.

Early morning mist, an important source of moisture, in the mountains above Kalk Bay (1). There are extensive cave systems (2) in the forests above Muizenberg and Kalk Bay, created when the mountains were folded into shape millions of years ago. Their sculptured shapes are matched by the twisted forms of trees (3) trying to get a grip on a rock-littered layer of humus.

2

3

Volcanoes on the Peninsula? Indeed. A huge fire on Devil's Peak at the start of the millennium is a reminder of the fragility of a seemingly robust environment.